discovermore

Forms of Energy

What Is Chemical Energy?

Virginia Elliott

Britannica Educational Publishing
IN ASSOCIATION WITH
THE ROSEN PUBLISHING GROUP

Published in 2026 by Britannica Educational Publishing (a trademark of Encyclopædia Britannica, Inc.) in association with The Rosen Publishing Group, Inc.
2544 Clinton Street, Buffalo, NY 14224

Distributed exclusively by Rosen Publishing.
To see additional Britannica Educational Publishing titles, go to rosenpublishing.com.

Editor: Danielle Haynes
Book Design: Michael Flynn

Photo Credits: Cover Bukhta Yurii/Shutterstock.com; (series background) Dai Yim/Shutterstock.com; p. 4 branislavpudar/Shutterstock.com; p. 5 fast-stock/Shutterstock.com; p. 6 guys_who_shoot/Shutterstock.com; p. 7 Smit/Shutterstock.com; p. 8 rktz/Shutterstock.com; p. 9 saran insawat/Shutterstock.com; p. 10 DKN0049/Shutterstock.com; p. 11 Shine Nucha/Shutterstock.com; p. 13 (top) Mark Agnor/Shutterstock.com; p. 13 (bottom) afotostock/Shutterstock.com; p. 14 Philip Pilosian/Shutterstock.com; p. 15 Rawpixel.com/Shutterstock.com; p. 16 GraphicsRF.com/Shutterstock.com; p. 17 Chendongshan/Shutterstock.com; p. 18 fantom_rd/Shutterstock.com; p. 19 Daniel Hoz/Shutterstock.com; p. 20 brgfx/Shutterstock.com; p. 21 Barou abdennaser/Shutterstock.com; p. 23 (top) Rimma Bondarenko/Shutterstock.com; p. 23 (bottom) Chatham172/Shutterstock.com; p. 24 Ilyashenko Oleksiy/Shutterstock.com; p. 25 Jabeen Fatima M J/Shutterstock.com; p. 27 (top) Jabeen Fatima M J/Shutterstock.com; p. 27 (bottom) Nandalal Sarkar/Shutterstock.com; p. 28 New Africa/Shutterstock.com; p. 29 Ozant/Shutterstock.com.

Cataloging-in-Publication Data

Names: Elliott, Virginia (Children's writer).
Title: Chemical energy / Virginia Elliott.
Description: New York : Britannica Educational Publishing, in association with Rosen Educational Services, 2026. | Series: Discover more: forms of energy | Includes glossary and index.
Identifiers: ISBN 9781641905602 (library bound) | ISBN 9781641905596 (pbk) | ISBN 9781641905619 (ebook)
Subjects: LCSH: Reactivity (Chemistry)--Juvenile literature. | Chemical kinetics--Juvenile literature.
Classification: LCC QD501.E45 2026 | DDC 541.394--dc23

Manufactured in China

CPSIA Compliance Information: Batch #QWBRIT26. For further information contact Rosen Publishing at 1-800-237-9932.

Contents

Energy Basics

Energy is the ability to do work. It's another word for power. It is one of the most basic ideas in science. Energy makes things move. It makes plants and animals grow, and it makes machines work. All activity in the universe can be explained in terms of energy and **matter**.

The sun provides energy to plants, which then give animals energy when they eat them.

Gasoline is a form of fossil fuel, which also includes coal and oil. These all are forms of chemical energy.

Energy is important, but it can be a bit hard to understand. That is because energy exists in many different forms. Some of the forms of energy are heat, light, electrical energy, chemical energy, mechanical energy, and sound energy. We see these forms of energy in our life every day. For example, the food we eat has chemical energy. Vehicles such as cars, trucks, buses, and motorcycles also run on the chemical energy stored in gasoline.

WORD WISE

ANYTHING THAT TAKES UP SPACE IS CALLED MATTER.

The different forms of energy can change into other forms. In a lightbulb, for example, electric energy changes to light and heat. When you burn a piece of wood in a campfire or a fireplace, the chemical energy of the wood changes to heat and light. Our bodies get the energy they need to grow, move, and think from the chemical energy stored in the foods we eat. Batteries also contain stored chemical energy. When you use batteries in a flashlight, the chemical energy from the batteries is changed into electricity and then into light.

The chemical energy stored in these pieces of wood changes into heat and light energy.

Solar panels change light energy from the sun into electrical energy to power other objects.

The fact that one form of energy can convert, or change, into another makes it easier for people to get the energy they need. If people need light, they can convert the chemical energy in candles into light energy. Or they can convert the electricity running to a lamp into light. This process of energy changing forms is called energy conversion.

Consider This

How might other forms of energy be converted into heat energy?

Elements and Chemical Compounds

All matter in the universe is made up of chemical elements. They are basic substances that cannot be broken down into simpler substances. Every known element has a symbol. The symbol for an element is sometimes the first letter of its name. H is the symbol for the element hydrogen and O is the symbol for oxygen.

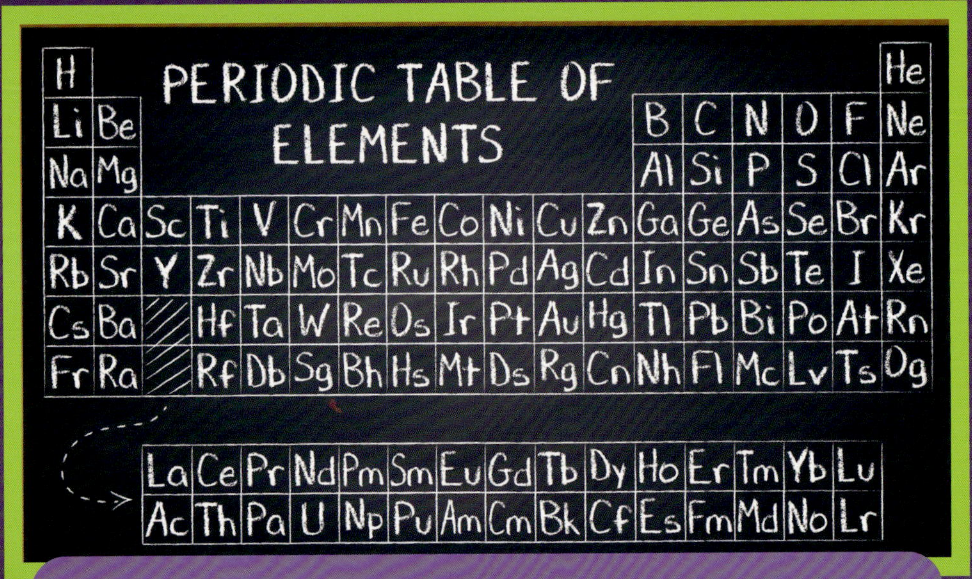

Scientists list all known chemical elements on a chart called the periodic table. Here we see the symbols for each element.

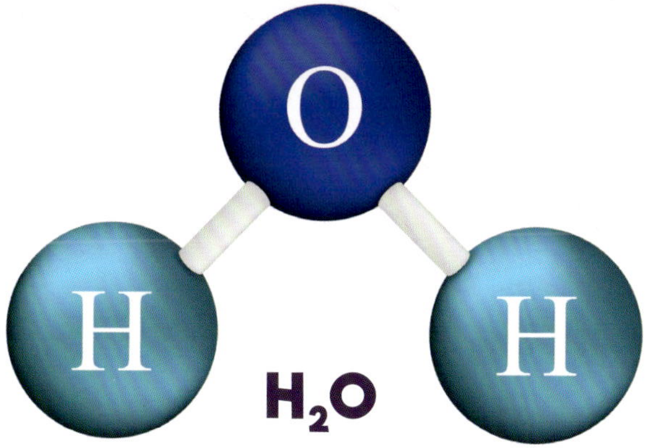

This diagram shows a water molecule, with two atoms of hydrogen (H) and one atom of oxygen (O).

The smallest unit of an element is called an atom. Atoms can be combined with other atoms to form molecules. For example, one oxygen atom and two hydrogen atoms combine to make a water molecule (H_2O).

Some substances are made up of molecules that contain just one type of chemical element. Oxygen gas (O_2) is made up of molecules of just oxygen atoms. Other substances are made up of molecules that contain atoms of two or more different chemical elements. Such substances are called chemical compounds. The most common chemical compound on Earth is silicon dioxide (SiO_2), which is one atom of silicon and two atoms of oxygen. You find this compound in rocks and sand.

compare and contrast

How are atoms and molecules similar? How are they different?

Chemical Reactions and Energy

Chemical energy is the reason the atoms in molecules hold together. All atoms are made up of smaller particles called electrons, protons, and neutrons. Strong chemical bonds are formed whenever atoms share electrons. Electrons are found alone or in pairs in the outer part of atoms. When two atoms with unpaired electrons get close, their unpaired electrons may form a pair. Both atoms then share the pair. This holds the atoms together.

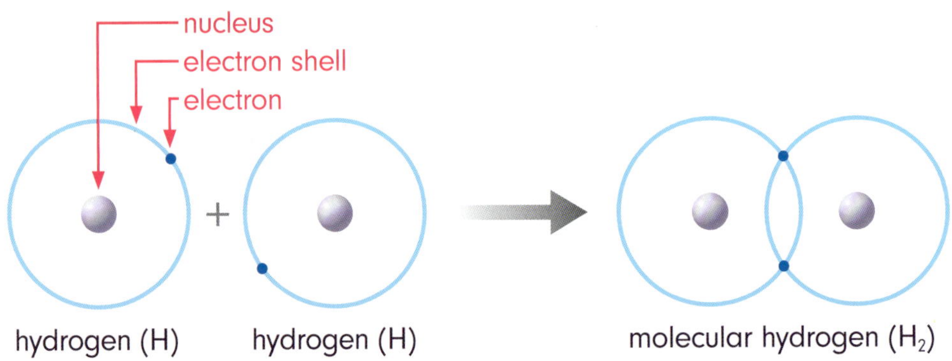

Electrons orbit or travel around the nucleus, or center, of atoms in what's called an electron shell. The more electrons an atom has, the more shells there are.

The chemical reaction that occurs when workers add water to dry cement causes the mixture of wet cement to then become warm as it releases energy.

When chemical bonds between atoms are formed or broken, chemical reactions occur. A chemical reaction is a process in which one or more substances are converted to one or more different substances. In a reaction, the atoms of the starting substances are rearranged, forming new substances with different **properties**. Energy plays a key role in chemical reactions. Breaking bonds requires energy, while forming bonds releases energy. Many times, chemical reactions release heat energy.

WORD WISE

PROPERTIES OF MATTER DESCRIBE EITHER A SUBSTANCE'S PHYSICAL FEATURES OR HOW THE SUBSTANCE CAN CHANGE INTO ANOTHER SUBSTANCE UNDER SPECIAL CONDITIONS.

Sometimes the release of heat energy during a chemical reaction can be dangerous. When certain chemicals are mixed to make dynamite, an explosion can occur if the person mixing them isn't careful. An explosion is a chemical reaction that gives off a large amount of energy. Many reactions give off energy without being explosive. In these types of reactions, the energy required to break bonds is less than the energy released as new bonds form. Heat energy is often released from these types of reactions.

In other types of chemical reactions, the energy required to break bonds is greater than the energy released as new bonds form. For such a reaction to occur, energy must be added to the process. Energy in the form of heat or electricity is often absorbed by these kinds of reactions. Any type of cooking done in a kitchen causes chemical reactions to occur in which energy, such as heat, is absorbed. The chemical reaction when you cook food on a grill is called the Maillard reaction.

The release of energy from dynamite is so great that it's often used in mining to break up rock.

compare and contrast

How are a burning candle and a frying egg similar? How are they different?

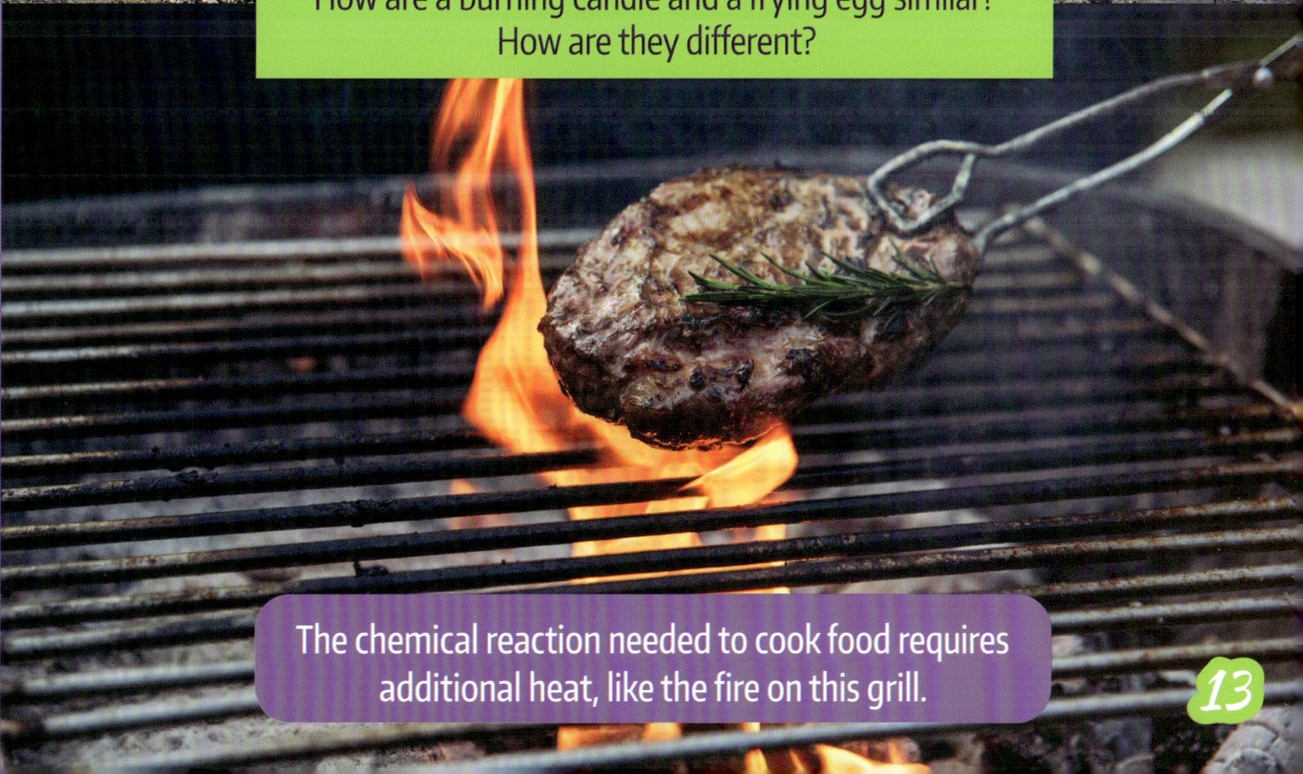

The chemical reaction needed to cook food requires additional heat, like the fire on this grill.

Get Moving

When you think of energy, you might think of cars speeding, people running, and motion. But that's not always the case. All forms of energy can be described as either potential energy or kinetic energy. Potential energy is stored energy that comes from an object's position or state. Chemical energy is a form of potential energy that is stored in the bonds of molecules. For example, potential energy is stored in a stretched-out rubber band or in the bonds of molecules that make up the gasoline inside a car's gas tank.

Chemical energy in the form of rocket fuel was stored in huge tanks attached to NASA's Space Shuttle. This potential energy helped the Space Shuttle fly 135 trips in its lifetime.

A chemical reaction turns the potential energy in fuel into kinetic energy in the form of heat energy.

If the person stretching the rubber band lets go, the rubber band will snap. If a driver starts a car engine, the gasoline is burned, which is a chemical reaction that releases a lot of energy—enough to drive the engine. As the rubber band and the car move, they gain kinetic energy. Kinetic energy is the energy of moving things. An object's potential energy can turn into kinetic energy. Kinetic relates to movement or motion.

Consider This

What is another everyday example of chemical energy changing into kinetic energy?

Energy and Plants

Energy stored in food is a type of chemical energy. Photosynthesis is the name of the process that plants, algae, and certain other organisms use to convert light energy from the sun into the chemical energy of food. This process takes place inside a body called a chloroplast. Photosynthesis is very important to life on Earth, because almost all living things depend on plants for food.

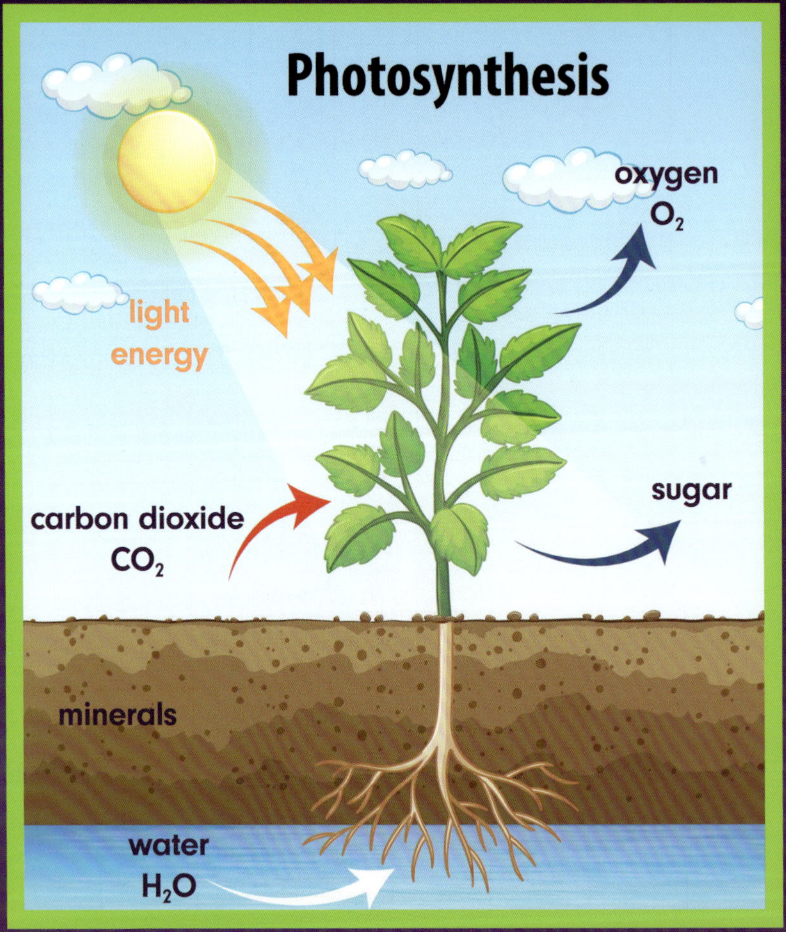

Chemical energy is stored in the sugars produced by plants.

Animals—and humans—get energy by eating plants, which have stored chemical energy.

Photosynthesis starts when a green substance in plants, called chlorophyll, absorbs energy from sunlight. Plants also take in water from the ground and carbon dioxide from the air. They use the light energy to change water and carbon dioxide into oxygen and sugars. The plants use some of the sugars and store the rest. The oxygen is released into the air. Animals that eat plants use the chemical energy stored in the sugars to get the energy they need to stay alive. The chemical energy created by photosynthesis is one of the most important steps of the food chain.

Consider This

Animals that eat only other animals are called carnivores. How do you think carnivores get chemical energy?

Energy from Food

Organisms use a process called cellular respiration to break down and use the energy from food. This chemical energy provides the power that **cells** need to work. Cellular respiration takes place in the cells of all living things, including animals, plants, fungi, and algae.

Both the grass and this cat use cellular respiration to make use of chemical energy from food.

The oxygen in the air we breathe breaks down the food we eat in cellular respiration.

Cellular respiration is a chemical reaction. During cellular respiration, a sugar called glucose is broken down in the presence of oxygen. This process releases chemical energy and produces carbon dioxide and water as waste products.

This is represented by the chemical equation **glucose** + **oxygen** → **chemical energy** + **carbon dioxide** + **water.** Glucose and oxygen (on the left of the arrow) are the reactants of the reaction. Chemical energy, carbon dioxide, and water (on the right side of the arrow) are the products of the reaction. Some organisms, such as yeasts, don't use oxygen and instead break down food using a process called fermentation.

WORD WISE

CELLS ARE THE SMALLEST UNITS WITH THE BASIC CHARACTERISTICS OF LIFE. SOME LIVING THINGS CONSIST OF ONLY ONE CELL. MOST PLANTS AND ANIMALS HAVE MANY BILLIONS OF CELLS.

Structures in cells called mitochondria have lots of different jobs in the body. One of the most important is cellular respiration. Mitochondria's main job is to produce chemical energy through cellular respiration. Some cells, such as those in the liver or muscles, have lots of mitochondria.

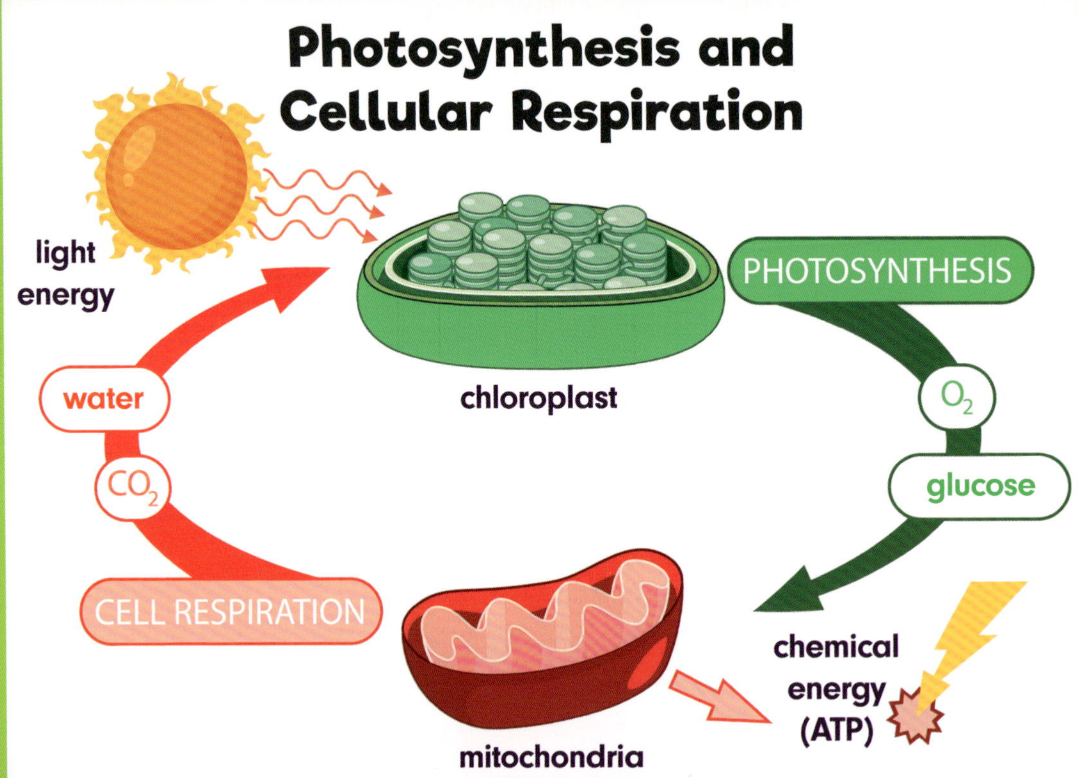

Photosynthesis creates oxygen (O_2) and food (glucose) for animals, who then, through cellular respiration, create carbon dioxide (CO_2) used by plants.

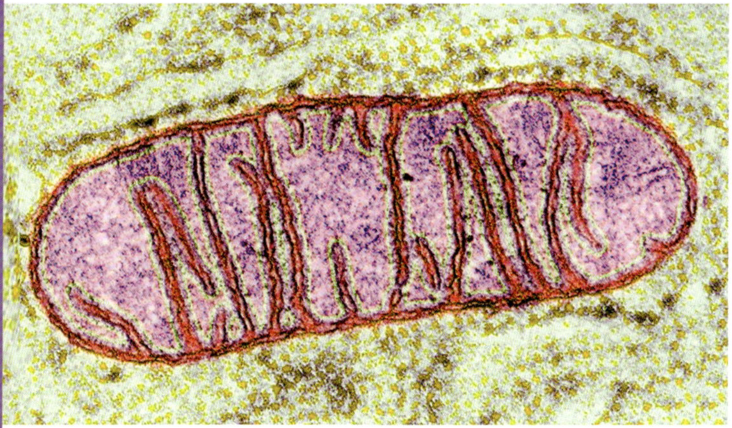

Mitochondria, which can be seen here under a microscope, produce adenosine triphosphate (ATP), store calcium, generate heat, and have a role in cell growth and death.

The chemical energy released in cellular respiration is captured in molecules called adenosine triphosphate, or ATP. ATP molecules leave the mitochondria and supply chemical energy to other places in the cell where it is needed. Cellular respiration also releases hydrogen and carbon. The hydrogen combines with oxygen to produce water, and the carbon combines with oxygen to produce carbon dioxide. The water and carbon dioxide that are formed during cellular respiration are released from the cell as waste products into the bloodstream. Carbon dioxide then goes into the lungs, where animals breathe it out.

Consider This

ATP molecules are chemical compounds. Once ATP reaches the parts of a cell that need energy, what do you think happens to the compound so that its energy is released?

Fermentation and Food

For the most part, cellular respiration needs oxygen to take place. Without oxygen, though, cells can still get chemical energy in a different way. This process is known as fermentation. Fermentation is a chemical change that happens in vegetable and animal substances. Fermentation often happens through the work of tiny living things called yeasts, bacteria, and mold. These living things create substances called enzymes. The enzymes break down food, such as glucose, into chemicals.

For thousands of years people have used fermentation to make bread, wine, beer, cheese, and other foods. When bakers add yeast to bread dough, the yeast breaks down the sugars in the dough. While this happens, carbon dioxide gas is released, making the bread rise. People eat many other fermented foods. Cheese and soy sauce are also made by fermentation. Scientists even use fermentation to make certain drugs, vaccines, and vitamins.

The bubbles you see in this sourdough starter, or dough, is caused by yeast fermentation.

compare and contrast

How are cellular respiration and fermentation similar? In what important ways are they different?

The fermentation in foods such as pickles, yogurt, and sauerkraut creates healthy bacteria for our bodies.

Chemical Energy in Batteries

Batteries aren't just a source of electrical energy. They also store chemical energy! Batteries give electric power to flashlights, radios, cell phones, handheld games, and other types of equipment. Batteries store chemical energy and convert it into electrical energy when they are used.

Inside a battery there are two pieces of metal in a liquid or a paste. These metal parts are called electrodes—one is called a cathode and the other is an anode. The liquid or paste, called an electrolyte, is a mix of chemicals. Each electrode has a point, called a terminal, that sticks out of the battery.

Terminals are the little knobs that stick out of the ends of batteries.

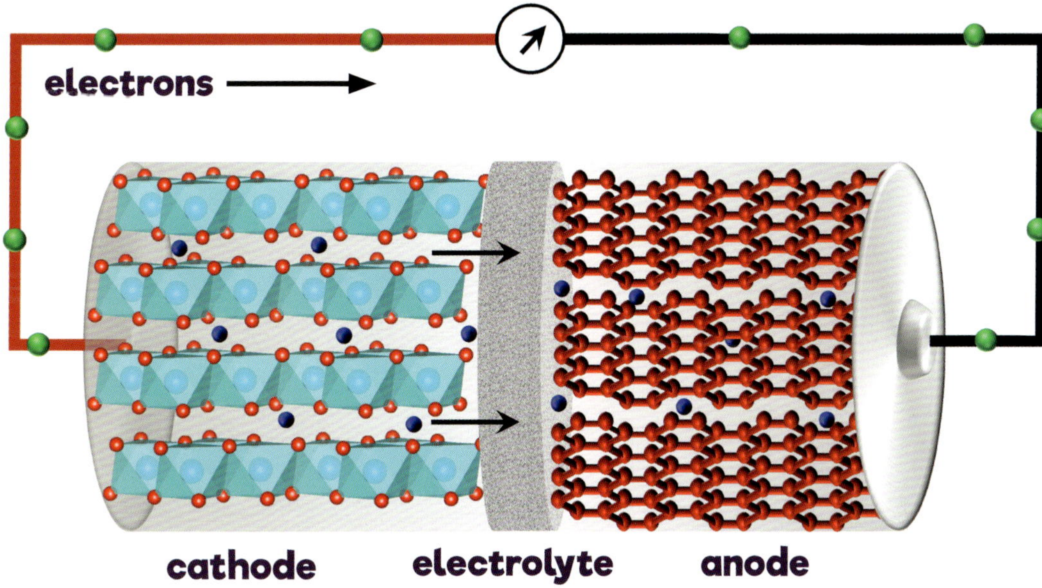

When you put a battery inside an electronic device, it causes a connection between the terminals, and electricity flows.

For a battery to work, the terminals must be linked by an outside wire. Then the chemicals in the electrolyte cause electrons to flow from one electrode to the other. A flow of electrons is an electric current. The electric current flowing through the wire is what makes flashlights and other electric equipment work. Batteries stop making electricity when the chemicals inside get used up.

Consider This

Some batteries can be recharged with an electric current from another source. How might this work?

Long-Lasting Fuel Cells

Fuel cells are like batteries in that they can convert chemical energy into electrical energy. Fuel cells, however, can supply electricity for a much longer period than batteries. They are used to provide power in large buildings such as hospitals.

A fuel cell is made up of a group of cells. Each of the cells is like a battery in that it has two electrodes and an electrolyte, which make an electric current. Fuel cells work differently than batteries by combining oxygen and fuel to produce electricity. Fuels such as hydrogen gas, methanol, or natural gas are broken down into hydrogen ions and electrons. Adding oxygen causes the electrons to move through an outside circuit.

The stored chemical energy in batteries eventually is used up. However, fuel cells never change. They continue to work for a much longer period of time because of a constant outside supply of fuel and oxygen.

Some new electric vehicles use hydrogen fuel cells for power.

compare and contrast

How are fuel cells and batteries similar? How are they different?

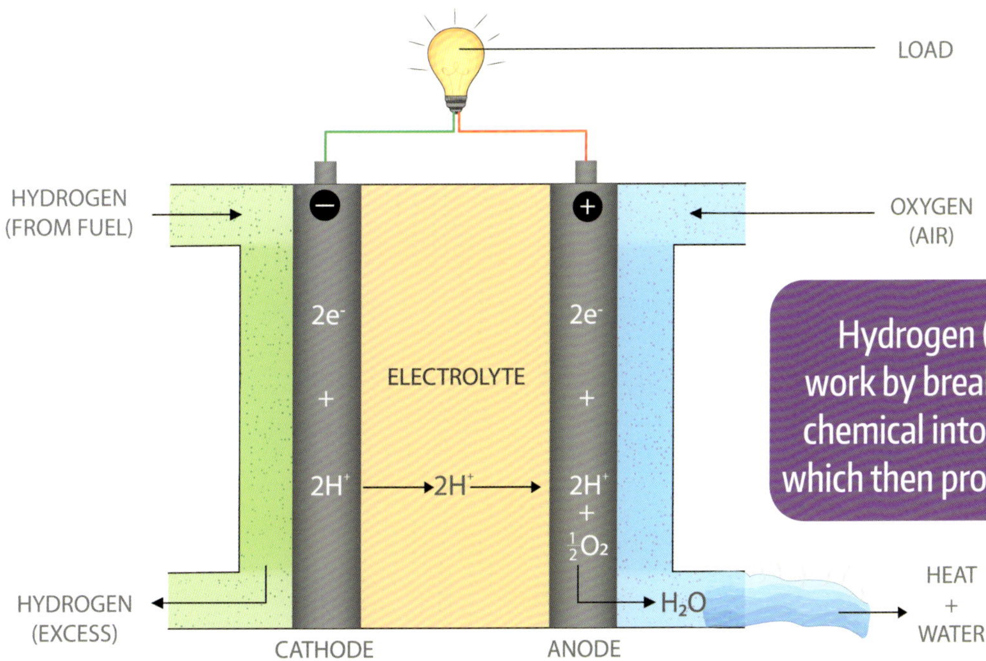

Hydrogen (H) fuel cells work by breaking down the chemical into electrons (e-), which then produce electricity.

Internal Combustion

Many cars, airplanes, and other types of vehicles use chemical energy transformed by internal-**combustion** engines. Internal-combustion engines burn fuel, such as gasoline, to do work. This process converts chemical energy into mechanical energy.

In the engine, moving parts called pistons slide up and down in cylinders. Valves at the top of the cylinders let in fuel and air and let out burned fuel. Spark plugs ignite the fuel.

Internal-combustion vehicles use fuels such as gasoline or diesel to run.

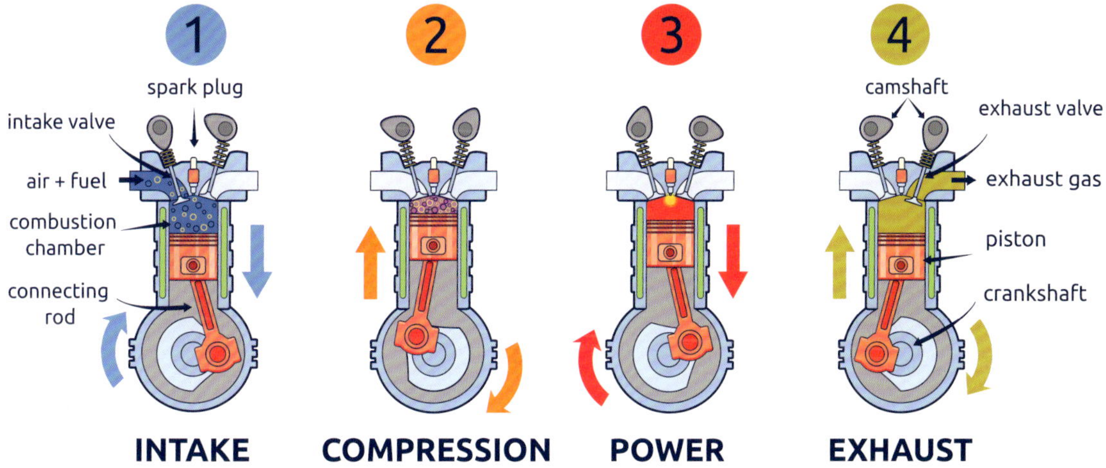

In 2023, about 77% of vehicles in the United States used internal-combustion engines.

As the engine runs, it goes through a cycle of movements. First valves open to fill the cylinders with fresh fuel and air. Next, the pistons squeeze, or compress, the fuel and air into a small space, creating pressure. At this point, the spark plug ignites the fuel. The explosion of the burning fuel moves the pistons down and up. The valves open as the pistons rise and push the exhaust gases produced by the explosion out of the cylinder. The moving pistons turn other parts of the engine to run the machine. When the engine runs out of gas, it stops and needs to be refueled.

WORD WISE

COMBUSTION IS A CHEMICAL PROCESS IN WHICH SUBSTANCES COMBINE WITH OXYGEN AND BURN FUEL.

Glossary

algae A group of living things that resemble plants (but are not plants), belong to a group called protists, and make their own food by photosynthesis.

basic Relating to something's essence.

carbon dioxide A gas that is produced when people and animals breathe out or when certain fuels are burned and that is used by plants for energy.

characteristic A trait or feature.

chemical A substance that is formed when two or more other substances act upon one another or that is used to produce a change in another substance.

dynamite A blasting explosive that is mostly made up of nitroglycerin absorbed in another substance.

electrical energy The energy produced when charged particles flow from place to place.

fungi A group of living things (including molds, mildews, and mushrooms) that are neither plants nor animals and that feed on other life forms, most often on dead plants and animals.

ignite To set on fire.

mechanical energy The total amount of energy an entire object has because of its motion and its position.

organism An individual living thing.

oxygen A gas found in air that is necessary for life.

particle A basic unit of matter and energy.

reactant A substance that enters into and is changed by a chemical reaction.

substance A physical material.

For More Information

Books

Hardyman, Robyn. *Nuclear Energy*. Bridgnorth, UK: Cheriton Children's Books, 2022.

O'Daly, Anne, and Lindsey Lowe. *The Basics of Chemical Reactions.* Buffalo, NY: Rosen Publishing, 2024.

Staats, Ann McCallum. *Energy Reactions in the Kitchen.* Huntington Beach, CA: Teacher Created Materials, 2025.

Websites

Energy Kids: Forms of Energy
www.eia.gov/kids/what-is-energy/forms-of-energy.php
This website from the U.S. Energy Information Administration explains more about the different forms of energy and how they're used.

Britannica Kids: Chemical Energy
www.kids.britannica.com/kids/article/chemical-energy/628739
Read more about chemical energy and look at diagrams on the Britannica Kids website.

Index